Cultural Celebrations

An Imprint of Pop!
popbooksonline.com

EARTH DAY

by Elizabeth Andrews

WELCOME TO DiscoverRoo!

This book is filled with videos, puzzles, games, and more! Scan the QR codes* while you read, or visit the website below to make this book pop.

popbooksonline.com/earth-day-dr

abdobooks.com

Published by Pop!, a division of ABDO, PO Box 398166, Minneapolis, Minnesota 55439. Copyright © 2024 by Abdo Consulting Group, Inc. International copyrights reserved in all countries. No part of this book may be reproduced in any form without written permission from the publisher. DiscoverRoo™ is a trademark and logo of Pop!.

Printed in the United States of America, North Mankato, Minnesota.

102023
012024

THIS BOOK CONTAINS RECYCLED MATERIALS

Cover Photo: Shutterstock Images
Interior Photos: Getty Images, Shutterstock Images
Editor: Emily Dreher
Series Designer: Colleen McLaren

Library of Congress Control Number: 2023939063

Publisher's Cataloging-in-Publication Data
Names: Andrews, Elizabeth, author.
Title: Earth Day / by Elizabeth Andrews
Description: Minneapolis, Minnesota : Pop!, 2024 | Series: Cultural celebrations | Includes online resources and index
Identifiers: ISBN 9781098245368 (lib. bdg.) | ISBN 9781098245924 (ebook)
Subjects: LCSH: Earth Day--Juvenile literature. | Environmental stewardship--Juvenile literature. | Holidays--Juvenile literature. | Cultural sociology--Juvenile literature.
Classification: DDC 394.26--dc23

*Scanning QR codes requires a web-enabled smart device with a QR code reader app and a camera.

TABLE OF CONTENTS

CHAPTER 1
Communities That Care 4

CHAPTER 2
History of Earth Day 10

CHAPTER 3
Celebrating Earth Day 18

CHAPTER 4
Going Green Together 24

Making Connections 30
Glossary 31
Index 32
Online Resources 32

CHAPTER 1

COMMUNITIES THAT CARE

Emilia sits at her kitchen counter as her dad prepares breakfast. He's making omelets with eggs from their backyard chickens. Mixed in is spinach and asparagus from their garden and cheese from the local farmers' market.

WATCH A VIDEO HERE!

Buying local ingredients or growing your own means food doesn't have to travel very far to make it to your plate. It is good for the environment!

It's important to wear gloves when picking up trash.

Today is Earth Day! Emilia's city is throwing a celebration. First, a group of community members will go for a walk and pick up litter. Then, everyone meets back at the park for music and food.

When Emilia gets to the park, booths are set up. Local businesses explain what they're doing to take care of the **environment**. Emilia's favorite booth is called Take Two, a secondhand shop. They are selling **vintage** T-shirts. She picks one out to buy. The person working the booth explains that buying secondhand clothing keeps unwanted clothes out of **landfills**. It's good for nature!

Buying secondhand means buying clothes or items someone else owned before you.

Being in nature is good for mental health.

Earth Day is a day to remember the importance of caring for the environment. It supports the protection of animals and **ecosystems**. Earth Day reminds people to ask their government leaders to pass laws that keep the earth healthy.

The word "climate" relates to weather patterns that exist in an area over a long period of time.

CHAPTER 2
HISTORY OF EARTH DAY

The first Earth Day was held on April 22, 1970. The year before, a large oil spill occurred in Santa Barbara, California. Oil spills spread thick, black liquid across the ocean surface. It covers sea animals and flows onto beaches. Many plants and animals die from oil spills.

LEARN MORE HERE!

Crowds of people gathered to hear speeches on the first Earth Day.

Hazardous waste teams clean up waste that may be harmful to nature or people.

People around the world were worried about troubles from changing technologies. One example was dangerous chemicals entering the **environment**. The oil spill happened because of human need for **fossil fuels**. People wondered what other problems might come about if rules to protect humans and the planet weren't made soon.

People have made signs for Earth Day for decades.

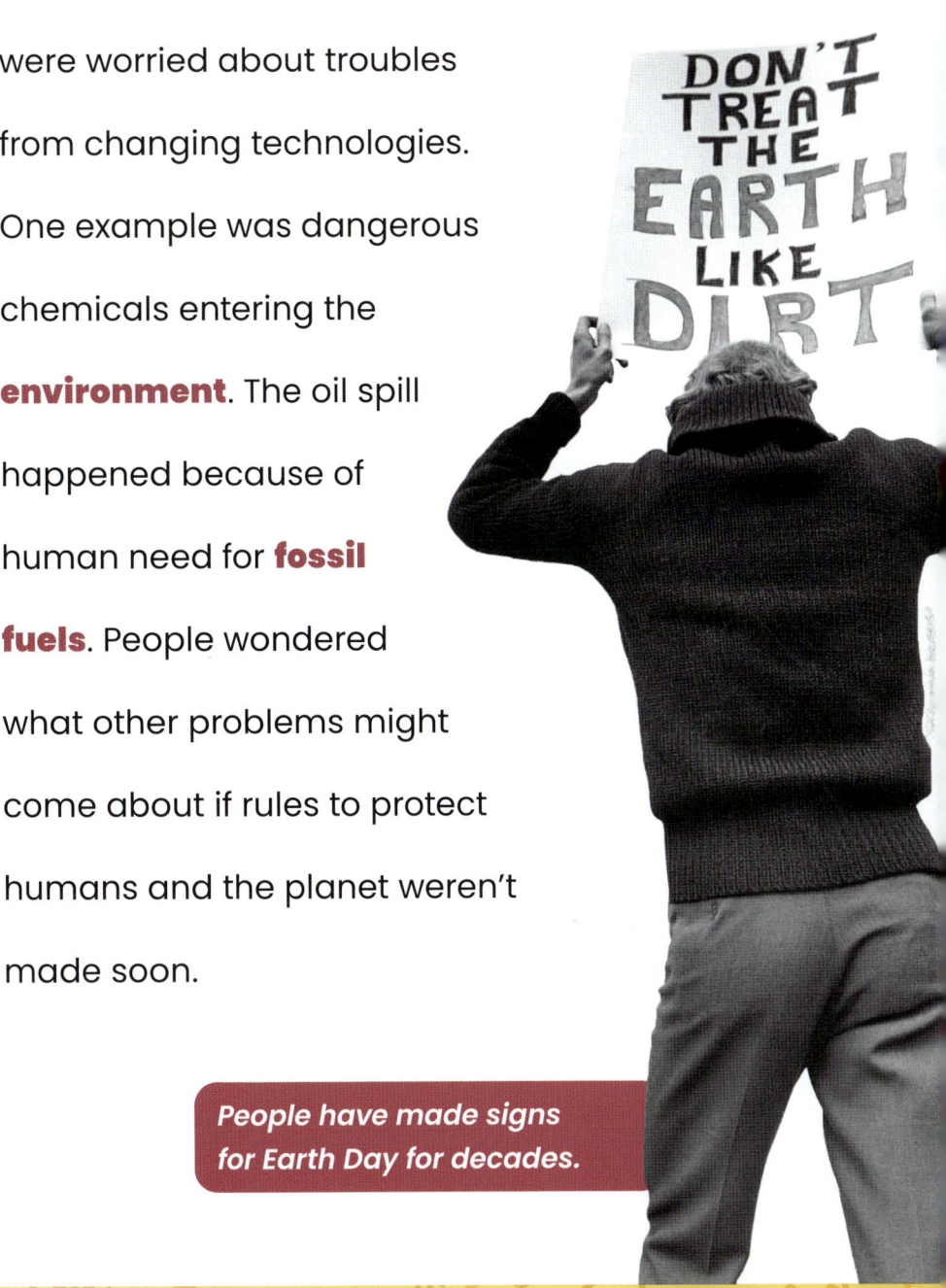

Gaylord Nelson grew up in Wisconsin. He loved the outdoors.

DID YOU KNOW? Conservation means the care and protection of air, minerals, plants, soil, water, and wildlife for future generations.

In 1970, Gaylord Nelson wanted to bring more attention to conservation. Nelson, along with a Harvard University student named Denis Hayes, created the first ever Earth Day. They held "teach-ins" to educate people about conservation. The biggest "teach-ins" were in Washington, DC, and New York City. Approximately 20 million people participated.

EPA AND MAKING LAWS

The Environmental Protection Agency (EPA) was founded after the first Earth Day. It was created to protect the environment and people from health risks. The Clean Air Act and the Endangered Species Act were passed soon after.

As the decades passed, the environment continued to suffer. In 1990, Hayes made Earth Day a global holiday. Now people all over the world can learn

Burning fossil fuels releases carbon dioxide into the air. This gas traps heat. It is bad for the planet.

about how to help the planet. People are focused on removing trash from the ocean and coming up with ways to use less fossil fuels.

CHAPTER 3

CELEBRATING EARTH DAY

Earth Day is celebrated by one billion people around the world. There are many ways to celebrate the holiday. People gather in large groups for marches and **rallies**. Sometimes the marches

EXPLORE LINKS HERE!

People care for the planet so future generations can enjoy it.

are to certain government buildings.

Marchers ask leaders to make new laws

or changes to protect the planet.

People may also march to natural landmarks that are affected by climate change. Lakes, rivers, and oceans face **pollution** that hurts the animals and plants living in them. Visiting those places on Earth Day helps spread awareness of the problems.

The ocean gets warmer as the planet does. The higher temperatures can hurt the animals that live there.

Community volunteering is a good way to make friends.

Community cleanups often take place during the week of Earth Day. One goal of Earth Day organizers is reminding people to care for their communities. People and businesses come together at community events. They can team up and plan ways to work together to go green!

WAYS TO GO GREEN

Going green means changing your lifestyle to do less harm to the **environment**. Another term for going green is "eco-friendly."

Bike or walk instead of driving.

Recycle.

Shop secondhand.

Use reusable bags and water bottles.

Use energy from solar and wind power.

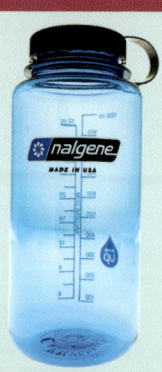

Earth Day is celebrated by 190 countries around the world. All events support caring for the environment. Earth Day reminds people that everyone on earth deserves clean air, water, soil, and food.

Choosing reusable water bottles instead of plastic makes less trash.

CHAPTER 4
GOING GREEN TOGETHER

The year 2020 marked the 50th anniversary of Earth Day. The theme of the day was climate action.

Earth's temperature is rising. As it gets hotter, plants and animals may not be able to survive. The warming is caused by human activities, such as **deforestation** and **fossil fuel** usage.

COMPLETE AN ACTIVITY HERE!

Many airplane engines use kerosene, a fuel made from oil. Companies work to create more fuel-efficient airplanes and greener engines.

Earth Day is a chance to teach people how to go green. In Barcelona, Spain, schoolchildren get to visit climate villages. They can participate in workshops led by forest police to learn

Planting trees helps remove carbon dioxide from the air.

It is best to view the northern lights on clear nights away from big cities.

about Spain's **ecosystems**. They can also play volleyball and do archery. There is a light show that looks like the aurora borealis, or northern lights.

An insect house provides shelter for helpful bugs in gardens. You can build one!

Austin, Texas, is the greenest city in the United States. Its city government works with community leaders, charities, and local businesses to throw an Earth Day **jubilee**. There is arts and crafts, food trucks, live music, and a butterfly zone.

In 2023, people in Hawaii came together to organize an ocean cleanup dive. Some 700 divers swam between the Hawaiian Islands gathering trash from the water and beaches. Every day new and creative ways to care for the **environment** are being created. Earth's future looks bright!

New inventions can help to solve pollution problems. This boat picks up trash in bodies of water.

MAKING CONNECTIONS

TEXT-TO-SELF

Do you do anything in your daily life to help the environment? Are there any other green changes you could make?

TEXT-TO-TEXT

Have you read any other books about Earth Day? If so, did you learn anything different from those books?

TEXT-TO-WORLD

People visit beautiful natural places on Earth Day. Where would you want to visit to celebrate Earth Day?

GLOSSARY

deforestation — the act of removing trees and clearing forests.

ecosystem — a community of organisms and their surroundings.

environment — the natural world, including air, water, land, and animals.

fossil fuel — a fuel formed in the earth from the remains of plants or animals. Coal, oil, and natural gas are fossil fuels.

jubilee — a celebration.

landfill — a place where lots of garbage is buried between layers of earth.

pollution — human-made waste that dirties or harms air, water, or land.

rally — a mass meeting intended to build group energy and excitement.

vintage — from the past.

INDEX

animals, 8, 10, 20, 24
Austin, Texas, 28

chemicals, 13
cleanups, 6, 17, 21, 29
community, 6–7, 21, 26–29

food, 4, 6, 23, 28
fossil fuels, 13, 17, 24

government, 9, 15, 19, 28

Hawaii, 29
Hayes, Denis, 15–16

marches, 18–20

Nelson, Gaylord, 15

ocean, 10, 17, 20, 29
oil spill, 10, 13

plants, 10, 20, 24

secondhand, 7, 22
Spain, 26–27

This book is filled with videos, puzzles, games, and more! Scan the QR codes* while you read, or visit the website below to make this book pop.

popbooksonline.com/earth-day-dr

*Scanning QR codes requires a web-enabled smart device with a QR code reader app and a camera.